W9-CGO-539

# SEA STARS

## MARINE LIFE

Lynn M. Stone

Rourke
Publishing LLC
Vero Beach, Florida 32964

www.rourkepublishing.com

PHOTO CREDITS: Title page, p. 6, 8, 12 (both), 13, 17, 21 (large), 22 © Marty Snyderman; p. 4, 7, 11, 14, 18, 21 (inset) © Lynn M. Stone

Title page: *A sea star wraps its arms around purple coral near the California coast.*

Editor: Frank Sloan

Cover and interior design by Nicola Stratford

**Library of Congress Cataloging-in-Publication Data**

Stone, Lynn M.
Sea stars / Lynn M. Stone.
p. cm. -- (Marine life)
Includes bibliographical references and index.
ISBN 1-59515-442-6 (hardcover)

Printed in the USA

CG/CG

Rourke Publishing
1-800-394-7055
www.rourkepublishing.com
sales@rourkepublishing.com
Post Office Box 3328, Vero Beach, FL 32964

# TABLE OF CONTENTS

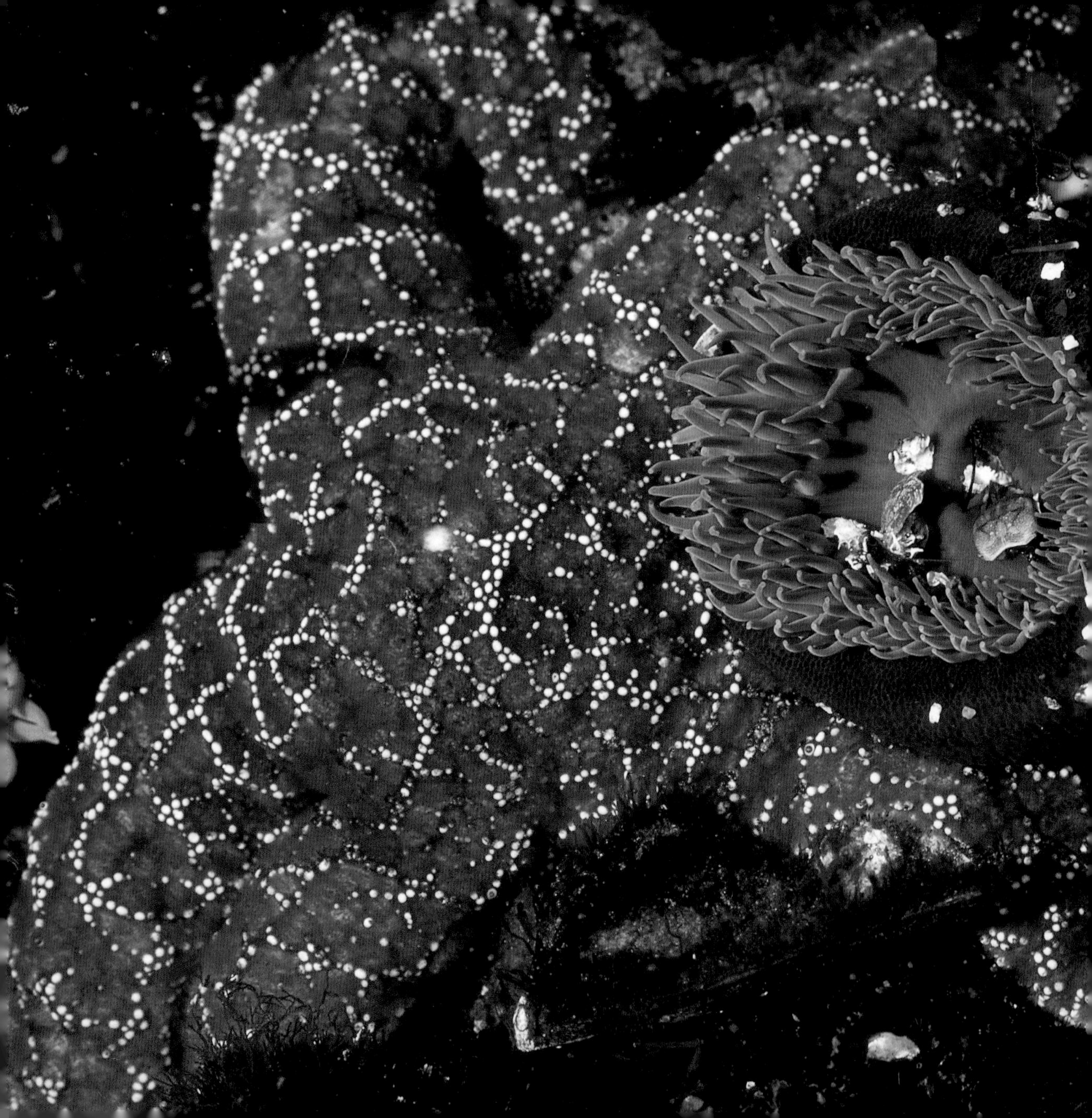

# SEA STARS

Sea stars are animals found in oceans and along ocean shores. They are also known as "starfish."

Sea stars are neither fish nor stars. Many of the most common **species**, however, are shaped like five-pointed stars.

Sea stars are **echinoderms**. Echinoderms are boneless **marine** animals with spiny skins and simple body plans. Sea cucumbers, sand dollars, and sea urchins are other common echinoderms.

*Many of the 1,800 species of sea stars have five arms.*

A sea star has neither head nor brain. It does have nerves, however, that sense light and probably sense food.

The center of most sea star bodies is a round disk. The animal's mouth is in the middle of the disk's underside.

*The underside of a sea star shows its mouth.*

*Sea stars and prickly green sea urchins are echinoderm cousins.*

*How many arms does this sunflower sea star have?*

A sea star's arms are its best-known features. The flexible arms spread from the disk like spokes on a wheel.

Sea stars can **regenerate**. If a sea star loses an arm, for example, it can grow a new one.

DID YOU KNOW?

Many species of sea stars have five arms. Others have four, six, eight, twelve, or even more than forty arms!

## What Sea Stars Look Like

Underneath each sea star arm are hundreds or even thousands of tiny tube feet with little suction cups. They allow sea stars to grip rocks and to catch **prey**. They also allow sea stars to move slowly across rocks or on the ocean bottom.

DID YOU KNOW?

A fast sea star may travel 10 feet (3 meters) in one hour.

*Sea stars use their arms to grip prey and hard surfaces such as rocks.*

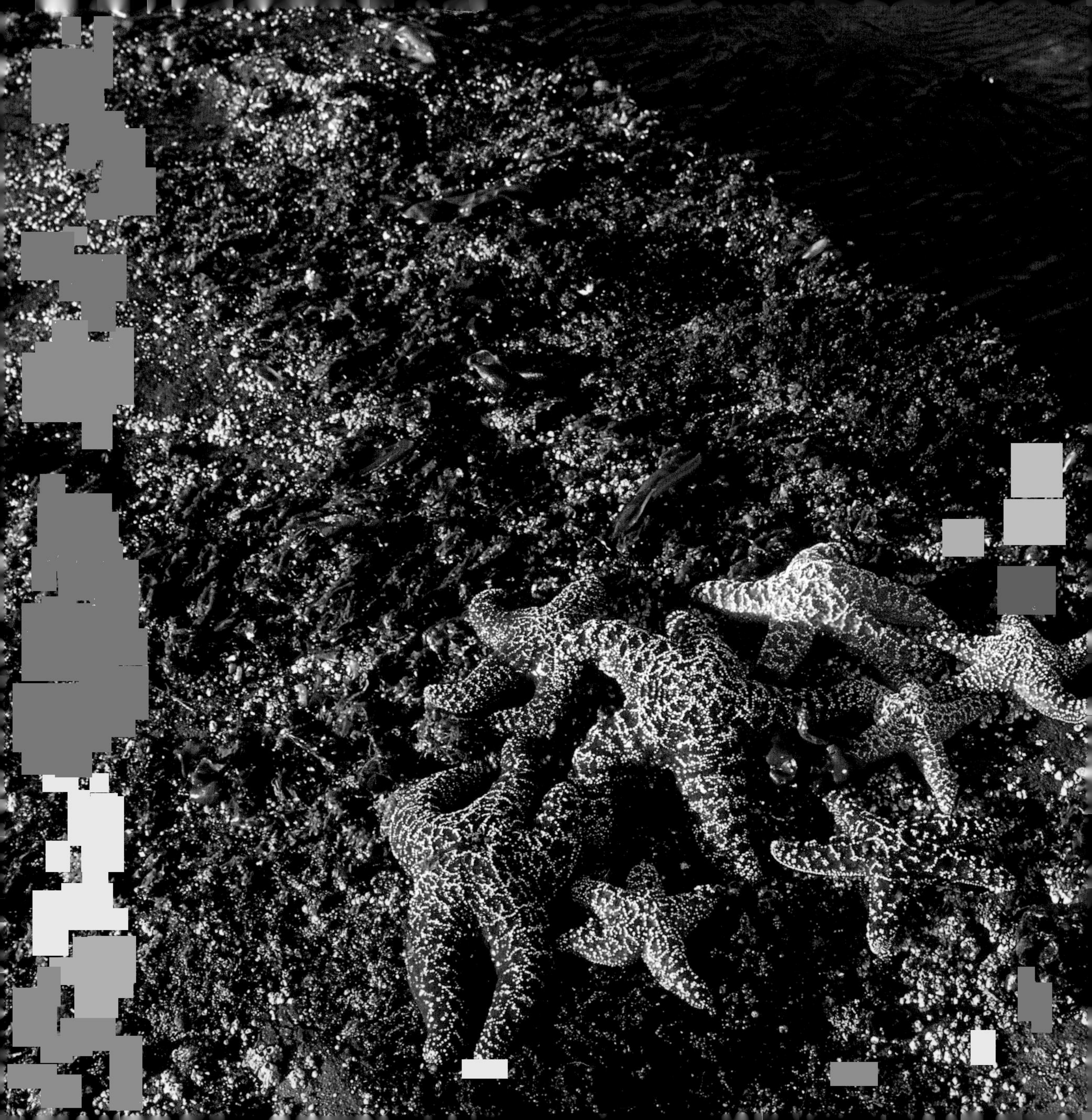

Sea stars come in a rainbow of colors.

Sea stars have rough skins, but many of them are not nearly as spiny as their cousins the sea urchins.

Sea stars may be yellow, orange, red, blue, purple, or other colors.

The most common sea stars in North America have fairly flat bodies and long arms.

*The basket sea star has an unusual shape.*

*The blood star is named for its red color.*

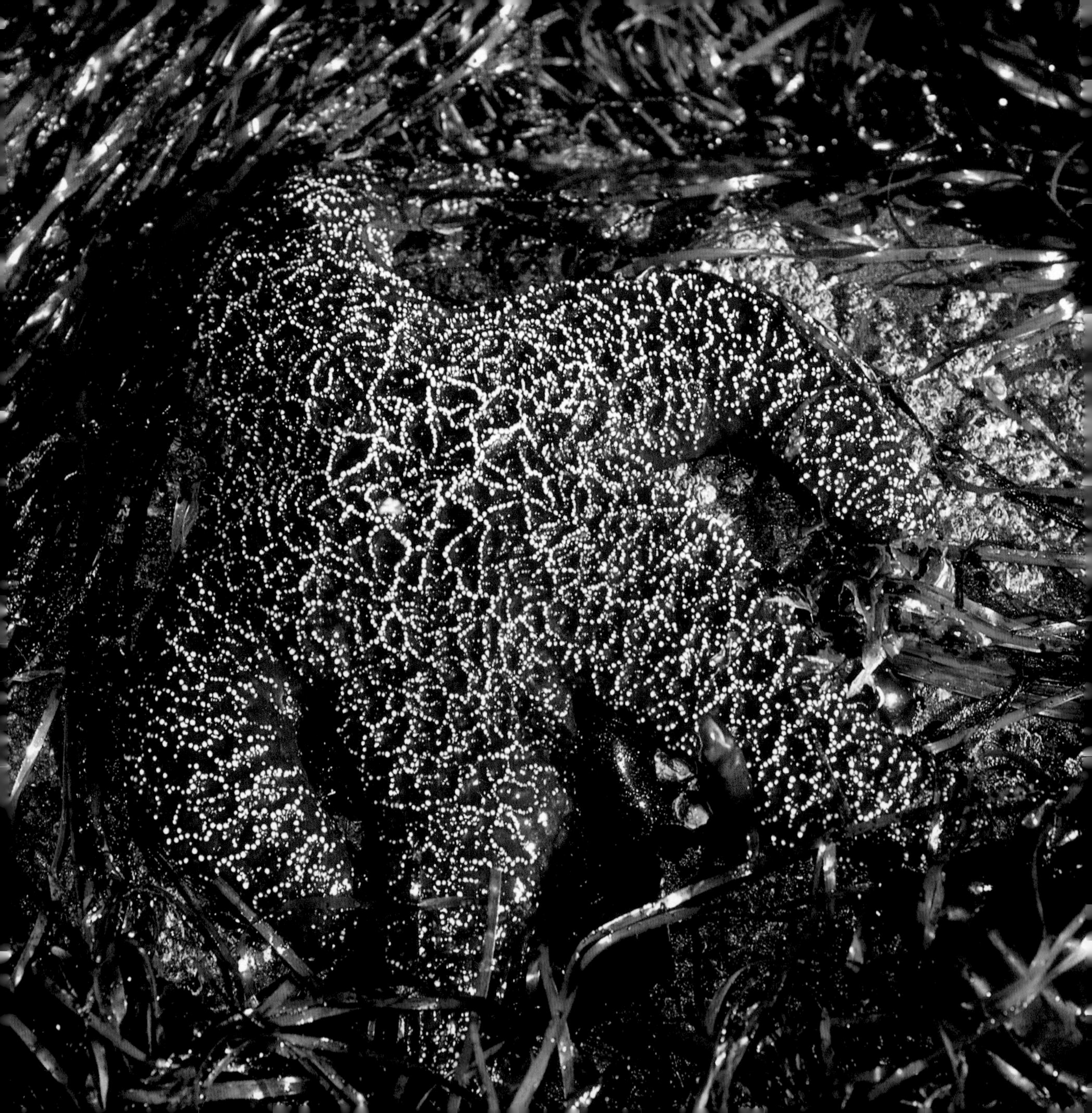

## Where Sea Stars Live

Sea stars are found in all the world's oceans, even in the icy cold Antarctic.

Many species live in shallow, coastal seas. They can be found quite easily during low tides.

*Sea stars can live out of water during low tides.*

## Predator and Prey

Sea stars are usually **predators**. What a sea star eats depends upon what kind of sea star it is and where it lives. It may eat snails, coral, dead animals, or tiny fish.

Many sea stars feed on shellfish such as oysters and mussels. A sea star opens a mussel by attaching its tube feet to the two closed halves of the mussel's shell. By steadily pulling, the sea star forces a crack between the shells.

*A spiny sea star feeds on a dead squid in coastal California.*

The sea star forces its stomach inside out through its mouth. It slides its stomach through the opening between the mussel's shells.

Once inside the shellfish, the sea star's stomach begins digesting the soft flesh.

## DID YOU KNOW?

Sea stars are sometimes prey for sea stars larger than themselves, marine turtles, and sea otters.

*Atlantic sea stars attack mussels in a Cape Cod, Massachusetts, tide pool.*

# The Life Cycle of Sea Stars

One way that sea stars make new sea stars is by laying eggs. The eggs hatch into tiny creatures that look nothing like adult sea stars. In some species, the baby sea stars, called **larvas**, can swim.

Swimming larvas live on tiny plants and animals that drift through the sea. After more growth stages, the larvas drop to the ocean floor to become adult sea stars.

*Bat stars engage in courtship behavior before laying their eggs.*

*A girl gently touches sea stars in an Oregon tide pool.*

# Sea Stars and People

Some people consider sea stars to be pests because they eat oysters and mussels. Oysters and mussels are popular sea foods for people as well as for sea stars. On the whole, sea stars are important members of marine communities.

*Crown-of-thorn sea stars sometimes destroy reefs popular with divers on the coast of Australia.*

# Glossary

**echinoderms** (eh KI nuh DURMZ) — any one of many marine invertebrates in the group including sea stars, sea biscuits, and sea urchins

**larvas** (LAR vuz) — early life stages in many animals, including sea stars

**marine** (muh REEN) — of the ocean

**predators** (PRED uh turz) — animals that hunt other animals for food

**prey** (PRAY) — any animal caught and eaten by another animal

**regenerate** (rih JEN uh RAYT) — to make a new organ or limb like the one that was lost

**species** (SPEE sheez) — one kind of animal within a group of closely related animals, such as an *Atlantic* sea star

# Index



## Further Reading


Hirschmann, Kris. *Sea Stars*. Thomson Gale, 2002
Svancara, Theresa. *Sea Stars and Other Echinoderms,* Volume 7. World Book, 2002


## Websites To Visit

http://cybersleuth-kids.com/sleuth/Science/Marine_Life/_starfish/index.htm
http://oceanlink.island.net/oinfo/biodiversity/seastars.html

## About The Author

Lynn M. Stone is the author and photographer of many children's books. Lynn is a former teacher who travels worldwide to pursue his varied interests.